BABIES AT THE ZOO

3

Elephant Calves

Susan H. Gray

Cherry Lake Press

Published in the United States of America by
Cherry Lake Publishing
2395 South Huron Parkway, Suite 200, Ann Arbor, MI 48104
www.cherrylakepublishing.com

Content Advisor: Dominique A. Didier, Professor of Biology, Millersville University
Reading Advisor: Marla Conn, MS, Ed, Literacy specialist, Read-Ability, Inc.

Photo credits: ©Susy Baels/Shutterstock.com, front cover; ©powell'sPoint/Shutterstock.com, 1, 2; ©Manfriday123/Shutterstock.com, 4; ©mikrus/Shutterstock.com, 6; ©Steffen Foerster/Shutterstock.com, 8; ©Villiers Steyn/Shutterstock.com, 10; ©Sabi Locksmith//Shutterstock.com, 12; ©LAMBERTO JESUS/Shutterstock.com, 14; ©Edwin Butter/Shutterstock.com, 16; ©T. Karanitsch/Shutterstock.com, 18; ©Natthawut Raruen/Shutterstock.com, 20;

Library of Congress Cataloging-in-Publication Data

Names: Gray, Susan H., author.
Title: Elephant calves / Susan H. Gray.
Description: Ann Arbor, Michigan : Cherry Lake Publishing, 2020. | Series: Babies at the zoo | Includes index. | Audience: Grades K-1.
Summary: "Read about how elephant calves grow to be the largest animal on land and how zookeepers take care of them. This level 3 guided reader book includes intriguing facts and adorable photos. Students will develop word recognition and reading skills while learning about how these baby animals learn and grow, what they eat, and how they socialize with each other. Book includes table of contents, glossary, index, author biographies, sidebars, and word list for home and school connection"—Provided by publisher.
Identifiers: LCCN 2019034089 (print) | LCCN 2019034090 (ebook) | ISBN 9781534158917 (hardcover) | ISBN 9781534161214 (paperback) | ISBN 9781534160064 (pdf) | ISBN 9781534162365 (ebook)
Subjects: LCSH: Elephants—Infancy—Juvenile literature. | Zoo animals—Infancy—Juvenile literature.
Classification: LCC QL737.P98 G71925 2020 (print) | LCC QL737.P98 (ebook) | DDC 599.6713/92—dc23
LC record available at https://lccn.loc.gov/2019034089
LC ebook record available at https://lccn.loc.gov/2019034090

Cherry Lake Publishing would like to acknowledge the work of the Partnership for 21st Century Learning, a Network of Battelle for Kids. Please visit http://www.battelleforkids.org/networks/p21 for more information.

Printed in the United States of America
Corporate Graphics

Table of Contents

About the Author

Susan H. Gray has a master's degree in zoology. She has written more than 150 reference books for children and especially loves writing about animals. Susan lives in Cabot, Arkansas, with her husband, Michael, and many pets.

Huge Babies

What is the biggest animal on land? The elephant!

Its head and body are huge. Its **trunk** is huge. Its legs are huge. Even the babies are huge!

What else may weigh
200 pounds (91 kilograms)?

A new elephant **calf** weighs over 200 pounds (91 kilograms). It already has teeth. Its legs are strong. The calf tries to stand soon after it is born.

The baby is **wobbly** at first. But it learns quickly. Soon, it is following its mother around. The calf stays with her and drinks her milk.

Later on, calves begin eating plants. Little elephants grow fast. In 3 years, a calf weighs as much as a car!

What a Job!

Some elephants live in zoos. **Zookeepers** have a huge job. Elephants eat hundreds of pounds of food each day. Zookeepers **fetch** them hay, fruits, and vegetables.

What part of this elephant would be difficult to wash?

Zookeepers also wash the elephants. They check their teeth. Zookeepers even **trim** the elephants' toenails.

Trunks and Trumpets

Calves love to play. They also like to swim on hot days. Tired little elephants take naps. Their moms and other females watch over them.

A lot happens in a young elephant's life. Calves learn to grab food with their trunks. They meet other elephants. They learn when to roar, **rumble**, and **trumpet**.

Elephants take about 12 years to grow up. Then they can have their own babies.

And those new elephant babies will be BIG!

Find Out More

BOOK

Hurt, Avery. *Elephants.* Washington, DC: National Geographic Readers, 2016.

WEBSITE

San Diego Zoo Kids—Baby Elephants
https://kids.sandiegozoo.org/videos/baby-elephants
Watch the video at this site to learn more about baby elephants.

Glossary

calf (KAF) the young of some animals, including the elephant
fetch (FECH) to go after and bring back something
rumble (RUHM-buhl) make a deep, rolling sound like thunder
trim (TRIM) to cut small pieces off
trumpet (TRUHM-pit) make a loud, powerful call
trunk (TRUHNGK) the body part an elephant uses to smell and to pick up things
wobbly (WOB-lee) shaky or not steady
zookeepers (ZOO-kee-purz) people who take care of zoo animals

Home and School Connection

Use this list of words from the book to help your child become a better reader. Word games and writing activities can help beginning readers reinforce literacy skills.

a
about
after
already
also
and
animal
are
around
as
at
babies
baby
be
begin
big
biggest
body
born
but
calf
calves
can
car
check
day
days
difficult
drinks
each
eat
eating
elephant
elephants
else
even
fast
females
fetch
first
following
food
fruits
grab
grow
happens
has
have
hay
head
her
hot
huge
hundreds
in
is
it
its
job
kilograms
land
later
learn
learns
legs
life
like
little
live
lot
love
may
meet
milk
moms
mother
much
naps
new
of
on
other
over
own
part
plants
play
pounds
quickly
roar
rumble
some
soon
stand
stays
strong
swim
take
teeth
the
their
them
then
they
this
those
tired
to
toenails
tries
trim
trumpet
trumpets
trunk
trunks
up
vegetables
wash
watch
weigh
weighs
what
when
will
with
wobbly
would
years
young
zookeepers
zoos

Fast Facts

Habitat: Grasslands and forests
Range: Africa and Asia
Average Height: 10 to 13 feet (3 to 4 meters) at the shoulder. African elephants are taller and heavier than Asian elephants.
Average Weight: 5,000 to 15,000 pounds (2,268 to 6,804 kilograms). Males can weigh twice as much as females.
Life Span: 50 to 70 years
Anatomy: The tip of the trunk has one or two flaps. These can do delicate, tricky tasks, such as picking up coins from the ground.
Behavior: Elephants hold tree branches in their trunks and use them to scratch their backs.

Index